KENT'S INDUSTRIAL HERITAGE
HERITAGE

James Preston

AMBERLEY

First published 2016

Amberley Publishing
The Hill, Stroud
Gloucestershire, GL5 4EP

www.amberleybooks.com

British Library Cataloguing in Publication Data.
A catalogue record for this book is available from the British Library.

ISBN 978 1 4456 6216 9 (print)
ISBN 978 1 4456 6217 6 (ebook)

Typesetting and Origination by Amberley Publishing.
Printed in Great Britain.

Contents

Introduction

Kent was described in *The Pickwick Papers* as a county of 'cherries, apples and hops', and has continued to be seen solely in the terms of the 'Garden of England'. However, although much of the county was, and is, purely agricultural, historically large swathes of Kent have supported important industrial activity. From the fourteenth to the eighteenth centuries the Weald was a centre for the production of woollens and iron. In the nineteenth and twentieth centuries, industry congregated in North Kent along the Thames, Medway and creeks of the Swale, and in urban centres such as Canterbury, Ashford and Dover, making Kent the most industrialised area of south-eastern England outside London.

Key to the economy of Kent was the county's geography and geology. Kent was strategically located on the Thames' approach to London, giving it easy access to the country's largest market. Demand from London, particularly after the Napoleonic Wars, was the main driver of the building materials industries and, as the centre of government, justice and commerce, was also a large consumer of high-class paper. With a long coastline, Kent was well placed for coastal trade and for trade with the Continent, while the Medway, Stour and the creeks of the Swale gave some access to the interior before the arrival of railways and improved roads. Cheap waterborne carriage was essential for particularly heavy bulk goods such as lime, cement, bricks and sand. The Darent, the Stour, and rivers and streams such as the Len and Loose provided water power for mills, which might alternate between fulling, corn, paper and gunpowder. In the Weald, the narrow valleys could be dammed to provide sufficient water to work blast-furnace bellows and forge hammers. Geology provided raw materials, including: iron deposits in the Weald for the iron industry; chalk from the North Downs for lime burning and cement; silica- and alumina-rich alluvial mud of the North Kent marshes and clay deposits for cement making and brickmaking; fuller's earth from Penenden Heath and Barming for processing woollen textiles; sand from Aylesford and Hollingbourne for the glass industry; and coal underlying East Kent, which was exploited in the twentieth century. The copperas industry, which utilised pebbles containing iron pyrites from the foreshore at Whitstable and Sheppey to produce ink for London stationers and dyes for the Wealden textile industry, was carried on at Whitstable, Gillingham, Queenborough and Deptford. Agriculture provided wool from sheep on the Downs and Romney Marsh for the woollen textile industry, linseed for oil-seed crushing and flax for linen, hides for tanning, and barley and hops for brewing. Woodlands supplied timber for shipbuilding and construction, as well as charcoal fuel and oak bark for tanning. London provided a source of linen rag for papermaking, centred on the mineral-free aquifer in the Maidstone area. Kent's location on the approach to London also gave it an important place in the defence of the realm. Apart from garrisons in strategic locations stimulating local demand – for example, for beer – Kent hosted two naval dockyards at Chatham and Sheerness, and supplied armaments in the form of cannon from the Weald, and gunpowder and explosives from Faversham, Dartford, Leigh and Cliffe. All this activity created a demand for means of transport, which was answered by barge-building yards springing up on the Thames, Medway and Swale, and for machinery, which was met by engineers in Dartford, the Medway Towns, Maidstone and Ashford. In the twentieth century, Kent's engineering sector was to see new products in the form of aircraft and road vehicles, including steam and petrol lorries, buses and motorcycles.

Now in the twenty-first century, with Britain in a post-industrial phase of its economic development, there is a danger that traces of past industry will be lost. Already most structures associated with the cement industry have been demolished, with a similar scenario for brick and paper making. The works of world-famous engineers have disappeared, along with breweries, maltings, boatyards, coal mines, gasworks and power stations. Urban industrial sites have become valuable for redevelopment as 'brownfield' sites.

This book attempts to record and act as a guide to important industrial sites. In addition to archives and written sources of information, the internet is a valuable resource. It is possible to have a virtual tour of sites using Google Earth, which now provides historical aerial photographs. Using 'Streetview', it is possible to inspect roadside structures, most of which have no public access. The 'Britain from Above' website has a limited selection of industrial sites accessible from the 'Aerofilms' photographic collection. The Kent Historic Environment Record website lists many of the industrial sites and gives access to the first to fourth editions of the 25-inch Ordnance Survey maps. Wikipedia also gives information about specific local industries. The internet provides a large numbers of sites relating to Kent industry, but care needs to be exercised over accuracy. Using these resources it is possible to determine which sites are worthy of a visit.

Chapter 1

The Textile Industries

The woollen textile industry became established at Cranbrook when it was settled by weavers from around Ghent after John Kemp was invited to move his business to England by Edward III in 1331. The area produced broadcloth – a fine, high-quality, luxury cloth of a minimum of 28 yards in length, 1¾ yards in width and 86 lbs in weight, much of which was exported via Blackwell Hall in London. By the early seventeenth century, in the region of 64 per cent of the Kent industry was concentrated in Cranbrook and the parishes of Biddenden, Benenden, Smarden, Sissinghurst, Hawkhurst, Staplehurst, Bethersden, Marden and Rolvenden, where production was dominated by major clothiers such as the Hendleys of Coursehorn and the Courthopes of Goddards Green, who worked on the 'putting out' system, employing hundreds of outworkers. The clothier bought in wool, which was washed and dyed on his premises, using woad, madder and copperas. Spinning was a cottage activity, which is estimated to have employed 6,000 – mostly part-time, poorly paid women who could spin 1 lb of wool per day. A known weight of spun wool was sent to weavers working around 650 looms, each of which required two men to operate. The woven cloth was sent to a fulling mill, after which it was was teased and trimmed by a shearman, before being collected at a cloth hall for forwarding to market in Maidstone or London.

Kersey weaving was contemporary with broadcloth production, but the cloth was 16 or 17 feet long, narrow, of lower quality and weighed 19 or 20 lbs. Production took place in the parishes to the north of the central Weald, including Smarden, Hothfield, Egerton, Lenham, Pluckley, Bethersden and Headcorn.

The 'New Draperies' developed in East Kent in the later sixteenth century, centred on Canterbury and Sandwich, and were boosted by the influx of Huguenot refugees in 1685. The textiles, including bays and says made from combed coarse wool, were lighter and finer than broadcloth; they included serge or worsted, which found a ready market in southern Europe.

The Kent woollen textile industry declined rapidly in the eighteenth century in the face of competition from Gloucestershire and particularly the West Riding of Yorkshire, where new technology was adopted and where old guild restrictive practices did not exist.

A silk-weaving industry flourished at Canterbury in the seventeenth century, largely carried on by French Protestant refugees. There were 340 looms in the town by 1710, but the number fell away till there were only six named silk manufacturers in 1784, by which time most weavers had migrated to Spitalfields.

Fulling to cleanse the cloth and thicken the pile was carried on in water mills, using fuller's earth dug from deposits at Penenden Heath and Barming. Fulling mills were particularly located around Maidstone on the Len and the Loose stream, where Lambarde said there had at one time been thirteen mills, with others around Cranbrook and Hawkhurst, on the Stour and the Darent. Fulling disappeared with the decay of the textile industry, and mills were converted to corn and paper mills.

The Old Studio, High Street, Cranbrook, is a fifteenth-century, Wealden-type cloth hall, now divided into three cottages. It has a characteristic tunnel through to the yard, where dyeing could take place. The High and Stone streets in Cranbrook are lined with timber-framed buildings associated with the textile industry, including the fourteenth-century George Hotel, 57 High Street, formerly a fifteenth-century wool hall, and the late fifteenth-century building housing Lloyds Bank. Many others were re-fronted in the eighteenth and nineteenth centuries.

Brass on the grave slab, in Cranbrook Church, of Thomas Sheffe, clothier, *c.* 1470–1520, who resided at Shepherds in the high street. His merchant mark is between the 'T' and 'S'.

Weavers' housing in Biddenden, dating from the late fifteenth century. The first floor was a continuous workshop for the accommodation of hand looms.

Weavers Court, Biddenden, which was believed to have been associated with the broadcloth industry. The central bay dates from *c.* 1535.

Biddenden cloth hall, of which the western section is fifteenth-century, while the east end is marked '1672' on the gable. The ground floor was used as cloth workers' workshops with living accommodation above, but the building was interconnected at attic level.

Weavers' cottages opposite Goudhurst church, dating from the seventeenth century and altered in the eighteenth century.

Reed's Cottages, Sissinghurst, built in the sixteenth century and altered in the eighteenth, were weavers' cottages.

Shakespeare House, Headcorn, is a late sixteenth-century Wealden cloth hall.

The fifteenth-century Smarden cloth hall with a hoist to the storage floors at the north end.

Weavers' houses, Sandwich.

The Huguenot silk weaver's workshop in St Peters Street, Canterbury, dating from 1500.

A sixteenth-century Huguenot weaver's house at 23 Palace Street, Canterbury, with a continuous first-floor mullion.

Chapter 2

Wealden Iron

The new blast-furnace technology, which had been transferred to Sussex by French immigrants, spread to Kent after 1550. Iron ore was available in the Weald from the Hastings beds (sand) and Wealden clay, where it occurred in nodules or tabular masses at a depth of 12 to 16 feet. Fuel in the form of charcoal was readily available, and water power to work bellows and forge hammers was provided by the Wealden streams that ran in fairly deep valleys, which could be dammed to provide a head of water. The Wealden iron industry flourished in the late sixteenth century when the main product was ordnance, cannon and iron shot. Fire backs, cast railings, gates and grave slabs were of secondary importance. Water-powered forges used hammers to remove carbon and impurities to produce wrought iron, which could be drawn into bars for nail making, horse shoes and similar products. There were at least fourteen furnaces and fourteen forges in the Kentish Weald, many of which had ceased production by the mid-seventeenth century. A small number of furnaces, including Barden, Tonbridge [TQ 548 425], Cowden and Lamberhurst [TQ 661 361], survived into the eighteenth century, but all had closed by the 1760s in the face of coke-fuelled iron production elsewhere.

No above-ground remains of the furnaces or forges exist, and only the sites of Chingley Forge, Goudhurst [TQ 682 335] and Furnace [TQ 685 327] (both now submerged in Bewl Reservoir), Scarlets Furnace, Cowden [TQ 443 401] and Biddenden Forge [TQ 822 383] have been excavated. Biddenden Forge operated from *c.* 1570 for a short time, with part of its 220-metre bay visible in the orchard adjacent to Hammermill Farm.

The most prominent above-ground traces of furnaces and forges are the pond bays (dams), which can be viewed at a number of locations. At Bedgebury Furnace, Cranbrook [TQ 7390 3470], the bay is 125 metres long and 3 to 4 metres high. The site can be approached via a bridle path. Frith Furnace, Hawkhurst [TQ 7360 3250], dating from the 1570s, has a bay 100 metres long and 3 metres high, which is now used as a farm track, and some trace of the loading area and wheel pit. It can be approached by a footpath and is some distance from a road. Bough Beach Furnace [TQ 482 476] has a 110-metre bay, breached in the centre by a stream, and can be reached by a public footpath. Postern Forge, Tonbridge [TQ 606 462], was operated by David Willard in 1553 and remained in use till *c.* 1600, where the 140-metre bay forms the bed of Postern Lane.

The pond and bay at Cowden Furnace [TQ 454 400]. Guns appear to have been cast here since at least 1574, with John Browne working the furnace between 1638 and 1664, and William Bowen, the gun founder, occupying the site possibly till as late as 1764.

The pond and 70-metre bay at Scarlet's Furnace, Cowden [TQ 443 401], where the furnace worked from at least 1590, producing guns and shot. It was still in production during the Dutch Wars in 1664.

The pond and bay at Horsmonden Furnace [TQ 695 412], where guns were cast from at least 1574, especially under Thomas Browne and successors up until 1668.

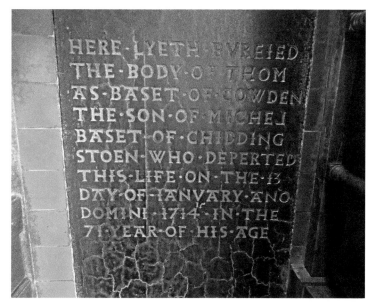

Iron grave slab in Chiddingstone Church, dated 1714.

Iron grave slabs at Cowden Church, dated 1710 and 1722.

Chapter 3

Gunpowder and Explosives

Kent was strategically located to be able to supply gunpowder to the key munitions stores in the Thames area from the sixteenth century. The county had streams and rivers to provide water power for the mills, and a safe means of transport for raw materials and the finished product. Relatively isolated locations in areas with trees to provide blast protection as well as the raw material for charcoal were to be found at Faversham, Dartford, Tovil and Leigh.

Powder mills at Faversham originated in *c.* 1558 on the stream between Ospringe and Davington, becoming known as the Home Works. Watermills for driving the incorporating process were introduced in 1733, work previously performed by horses. The importance of the works is underlined by their purchase in 1759/60 by the government as part of the Royal Gunpowder Mills. An explosion in 1781 resulted in the opening in 1786 of the Marsh Works [TR 013 626] in a more remote location outside the town. Both works were operated by John Hall of Dartford from 1815 (Home) and 1832 (Marsh) until 1898, when they were merged into Curtis & Harvey, eventually being closed by successor firm ICI in 1934.

The Oare gunpowder works [TR 0020 6228] were in operation by 1719. They came into the possession of Pigou & Andrews, who ran the Dartford powder mills in 1798 and were sold to John Hall in 1812. After becoming part of Curtis & Harvey, the works eventually passed into the hands of ICI Ltd to be closed and largely demolished in 1934. The site, to which there is public access and parking, has a visitor centre, which is open at weekends.

A site by the Swale at Uplees [TQ 999 650] was chosen in 1873 by the Cotton Powder Company to make guncotton, which was first produced at the Marsh Works in 1846 but discontinued after an explosion in 1847. 'Tonite', a mixture of guncotton and barium nitrate, was added to the range in 1874, followed by nitroglycerine in 1892 and cordite in 1896. In 1913, the Explosives Loading Company opened a factory adjacent to the west, which was the site of a massive explosion in 1916. The sites were closed by Explosives Trades Ltd in 1919.

Powder mills were set up on the Darent at Dartford in 1732 by Pike & Edsall on the site of John Spilman's paper mill. They were operated from 1788 by Andrews & Pigou, and from 1898 by Curtis & Harvey. At its peak, the site covered 50 acres with around 100 buildings. The mills ran until 1909.

The site near Leigh of a gunpowder mill, listed in *Bailey's British Directory* of 1784 as being run by Thomas Hooker & Co., has not been located. However, gunpowder mills were erected at Leigh [TQ 5710 4655] in 1811 by a partnership that included Sir Humphrey Davey. Power was provided by a mill race using a channel dug from the Medway above the site, with the tail race a parallel channel dug to empty downstream, leaving the works buildings on a narrow tongue of land between. The works were bought by the Curtis family in 1859, becoming Curtis & Harvey in 1898. Latterly the mills concentrated on smokeless sporting powders. The east of the site was later used for small-scale chemical explosives' manufacture. The works, which were closed by ICI Ltd in 1934 and demolished to ground level, are on private land; there is no access.

In 1892 Hay Merricks & Co. set up a small gunpowder storage and packing facility at Lower Hope Point [TQ 7300 7850] on the Thames, to the north of Cliffe village. The store was to the west of the current site, which extends over 128 hectares. The firm merged into Curtis & Harvey Ltd, along with seven other firms in 1898, after which the site was used for the production of chemical explosives. From 1901 onwards, the range of products included gun cotton, gelatine explosives, dynamite and cordite. The site was government controlled between 1914 and 1918, with rapid expansion of output from 1916 when the new H. M. Cordite Factory was developed on the east of the site. Post-war, the demand for explosives sharply declined, with the result that the site was closed in 1920, and substantially demolished by 1922. On site there are concrete foundations and the walls of numerous 'danger buildings', many within earth blast mounds. As there is no public access, the extensive remains can be seen from the sea wall or the public right of way to the west.

Chart Mills, Faversham [TR 0096 6124], are all that survives of the Home Works. Both pairs of incorporating mills were rebuilt in 1815 and were restored to their present condition, after having been partly demolished in 1934.

The reconstructed incorporating mill at Chart Mills, with edge runners salvaged from Oare.

The beds of water-powered incorporating mills at Oare, where the building, as with all danger buildings, would have been lightly constructed in timber.

Part of the canal network at Oare, which was used for moving gunpowder and materials by punt as well as to power the mills.

The corning house at Oare. The machinery was controlled from behind the blast wall, which had been rebuilt after an explosion. This building would have been of insubstantial timber, with power originally drawn from a waterwheel situated to the right of the photograph.

The work space behind the blast wall for workers controlling the corning-house machinery.

Incorporating mills, built in 1925 when the Home, Marsh and Oare Works were remodelled and modernised. Each mill was subsequently driven by an electric motor, housed beneath the floor of each unit.

Under-driven incorporating mill, of the type used at Oare after 1925. Note that the process was not grinding, but rolling the ingredients together – hence the space between the trough and edge runners.

The barge dock at the Cotton Powder Company, Uplees.

The jetty for the Explosives Loading Company, which was linked to the works by a tramway.

The concrete floors of buildings, the walls of danger buildings with surrounding earth blast banks and other features can be seen over a wide area at Uplees. The site is now in the hands of the RSPB as a bird reserve, and is sometimes partially flooded. The site is best seen from the Swale river wall.

Remains of an incorporating mill off the Darent Valley Path, Powdermill Lane, Dartford [TQ 5478 7282], where two pairs of mills are visible.

A pair of incorporating mills at Leigh.

The largest buildings at Cliffe, dating from 1914–18.

Remains at Cliffe, associated with cordite manufacture.

Chapter 4

Boat and Ship Building

Barges provided a cheap means of transport for Kent's bulky products, such as bricks, cement and agricultural products, including straw and hay. They also brought return cargoes of coke from the London gasworks for the cement works; 'rough stuff', which was ash from dust contractors, for the brickworks; rag for papermaking; and manure from the London stables. To meet demand, barge-building yards were set up along the Thames and Medway and the creeks of the Swale, requiring no permanent buildings and little fixed equipment as barges could be built on any firm shore. Bawley boats for fishing and oyster dredging were also built. Boatbuilders were supported by sail makers, and engineers such as Taylor & Neates of Rochester, who made winches and barge gear.

On a larger scale was Pollock's yard, Faversham [TQ 0195 6201], which was operational between 1875 and 1970. Steel ships were built here, including TIDs and landing craft, which were launched sideways. All that remains are some ruinous sheds behind a housing development.

Chatham dockyard was Kent's biggest 'industrial' site, producing ships and boats for the Royal Navy. Its buildings date from the eighteenth and nineteenth centuries, notably the covered slips, which were constructed at the end of the wooden-walled warship era. These were largely to prevent rot setting in as a result of rainwater collecting in partially constructed hulls. Also of note is the ropery, which houses some working machinery from 1811. Sheerness Dockyard [TQ 9087 7529] is a working port and not easily accessible – it contains the pioneering iron-framed boat store.

Hollow Shore barge-building and repair yard [TQ 0170 6363], which operated from the mid-nineteenth century at the confluence of the Oare and Faversham creeks on the busy barge routes from brick, cement and gunpowder works.

The site of the slip at the Crescent yard, Frindsbury [TQ 745 694], was probably in use as a warship yard during the Napoleonic Wars. It became a barge yard operated by Gills, and latterly a repair yard for the Rochester & London Trading Company (Crescent Shipping).

Sail loft and carpentry shop of Goldfinch's Yard, Island Wall, Whitstable [TR 1018 6623], which built barges and fishing boats. Bought by Anderson, Rigden & Perkins in 1917, the last wooden barge was launched in 1924, after which motor boats and yachts were built.

No. 3 slip, Chatham Dockyard [TQ 759 695], was constructed in timber in 1837, the third and only surviving example at Chatham of a covered slip with a roof designed by Robert Seppings. The slip is 300 feet long, 145 feet wide and 62 feet high, and provides sufficient headroom for raising masts.

The roof timbers of No. 3 slip.

Covered slips Nos 4, 5 and 6 at Chatham Dockyard were built in iron, with corrugated iron roofs, to the design of Capt. Mould of the Royal Engineers, and erected by Baker & Son. The slips are pioneering, long-span buildings, which give a headroom of 66 feet.

No. 7 slip, Chatham Dockyard, is the last covered slip of the wooden-wall era. Designed in 1851 by Col. G. T. Greene of the Royal Engineers and constructed in 1853, it had provision for overhead travelling cranes.

Mast houses and mould loft at Chatham dockyard [TQ 7602 6942], completed between 1753 and 1755.

Chapter 5

Paper Making

The first white paper mill in Kent was set up at Dartford in 1588 by the Spilman family on the site of a former iron-slitting mill, remaining in operation until *c.* 1724. The industry spread particularly to the Len and Loose streams, the Darent, and the Dour, where there were pre-existing fulling or corn mills, which could be easily converted to work beaters for rendering rags to fibre. Initially paper was handmade using a deckle (wire mould) dipped into a vat of rag fibres. Along with white paper, brown and wrapping papers were made from locally available discarded rope and sails. Kent was at the forefront of technical innovation, with James Whatman of Turkey Mill introducing moulds of woven wire mesh in 1756; the Fourdrinier machine being first used in Kent at River Mill, Dover, in 1807; Brian Donkin at Halls of Dartford developing papermaking machines; Townsend Hook at Snodland experimenting with straw paper; and the use of wood pulp being developed by C. D. Ekman and used at Northfleet in 1881. The first steam-powered mill was built at Buckland, Dover, in 1834, and the industry became largely steam powered in the second half of the nineteenth century. The use of steam enabled the industry to migrate to Thames-side locations at Dartford and Northfleet, where wood pulp could be unloaded from sea-going vessels. During the late twentieth century, there was a rapid decline in paper production in Kent, with most of the more modern, urbanised mills closing and being demolished, leaving the earlier rural mills to represent the industry.

Turkey Mill, Ashford Road, Maidstone [TQ 7721 5950], was converted from a fulling mill in the late seventeenth century. Initially under George Gill and later under James Whatman, it produced fine white paper – in 1782, it accounted for about 8 per cent of national output. The Hollingsworth family, who had operated seven mills with thirty-two vats around Maidstone in 1830, concentrated production at Turkey Mill after introducing a Fourdrinier machine in 1846, and adding two more by 1858, with their other small mills being closed. At one time powered by three overshot wheels, water power was last used in the early twentieth century. The drying sheds that date from *c.* 1740, the mill house and chimney remain converted for commercial use.

Horton Kirby Paper Mill [TQ 5631 6952] operated as a water-powered flour mill until *c.* 1830, when John Hall of Dartford, maker of paper machines, converted the buildings for paper making. The mill was converted to steam power in the 1850s. The buildings were replaced in 1880/81 and three new paper machines installed. The mill ceased production in 2003. There were also former paper mills on the Darent – at Shoreham, Eynsford, Sundridge, Hawley and Dartford.

The Italianate chimney at Horton Kirby, dated 1881.

Ford Mill, Little Chart [TQ 940 460], was working by 1776. In the nineteenth century, high-quality, handmade papers were produced. Production of paper ended in 1944, after which flong for typesetting moulds in the printing industry was produced. Apart from the mill pond, sluice gate and wheel pit, little remains of the original watermill.

Little Ivy Mill, Loose [TQ 755 525], on the Loose Stream was the site of a paper mill from the 1650s. Rebuilt as a corn mill in 1856, it worked until 1912, after which the wheel was removed. It is now a private residence.

Hayle Mill, Tovil [TQ 755 538], on the Loose Stream was the last mill in Kent to produce handmade paper. Built as a water-powered mill in *c*. 1808, steam engines were introduced in the late nineteenth century to run the beaters. Rag sorting and vats were housed in the front building, with the drying shed behind. The site has been converted for residential use.

Townsend Hook Paper Mill, Snodland [TQ 7075 6172]. Paper was made on this site from *c.* 1700. The mill was converted to steam power in the 1830s, and burnt down and rebuilt in 1906. The oldest current buildings date from the 1930s.

Buckland Paper Mill, Dover [TR 304 428]. Originally a flour mill on the Dour, it was converted to papermaking by 1638. The current buildings were erected after a fire in 1887. The mill was closed by Arjo Wiggins in 2000.

Chartham Paper Mill [TR 1080 5498] was originally built on the site of a fulling mill and converted to paper in 1730. It was here that tracing paper was invented. The mill, which was rebuilt and opened in 1949, is still the major supplier of tracing paper in Europe.

Associated with the paper industry at Snodland was Nichol's paper bag factory [TQ 7059 6161]. The factory site was first used *c.* 1855 for the manufacture of silk, but the venture failed. It then became a steam-powered bag factory, and remained active until 1981.

Chapter 6

Lime and Cement

The production of lime on a commercial scale developed on the Thames and Medway with the expansion of London after the Napoleonic Wars. Lime-works sites changed to the production of Portland cement in the second half of the nineteenth century and therefore kilns and equipment were destroyed. Over sixty cement works have operated in Kent, all located between Dartford and Faversham, apart from one short-lived venture at Folkestone. All have been demolished, mostly to ground level, leaving few remains. The most visible features left by the cement industry are the quarries, many of which are now being redeveloped for residential use as at Holborough and Eastern Quarry, Northfleet. The most complete kiln is a bottle kiln, which is said to date from 1845, on the site of William Aspdin's Northfleet works [TQ 6171 7500], but there is no public access.

The buried base of a bank of limekilns, dating from the 1840s, on the site of Lees Lime Works, Halling [TQ 7080 6317].

Base of a bank of limekilns in the White Pit, Halling [TQ 6948 6497], operated by Formby Bros in the second half of the nineteenth century and abandoned after 1907 when the Batchelor Bros purchased the site for cement production. Access via Halling Parish Council.

Limekilns alongside the Pilgrims Way, Charing [TQ 9681 4904], last used *c.* 1940. It was photographed here in the 1990s, but is now overgrown and in a ruinous state.

Nine bottle kilns were linked to a tall iron chimney, sections of which are still on site at Francis & Co., Cliffe [TQ 7096 7688].

Base of an edge-runner clinker grinder at Cliffe Creek. The edge-runner stones are used as parish boundary markers at Cliffe Woods and Cliffe, and were also probably used to grind chalk when the site became a whiting works in around 1900.

Base of a chamber kiln at Cliffe Quarry [TQ 7263 7640], which is almost certainly Isaac Johnson's prototype of 1878.

Square concrete columns at the West Kent Cement Works, Burham [TQ 7133 6236], supported an elevated tramway that brought chalk from Margett's Pit. Cement was produced from the 1870, with a new plant for quick-drying cement added in the early twentieth century.

The chalk wash mill at British Standard Cement Co. quarry, Berengrave, Rainham [TQ 8208 6712]. It dates from 1912, when the cement works were constructed.

Berengrave wash mill collecting tank is in the foreground, with the engine house behind.

The clay wash mill, from which clay dredged from Cliffe Lagoons [TQ 7215 1771] was pumped to vessels on the Thames for transportation to Northfleet and Swanscombe.

Remains of a slurry-drying back, Burham Cement Works [TQ 7210 1613].

Cliffe Lagoons were dredged for clay to mix with chalk in cement making. Clay was also obtained from the Medway marshes and dug at Shorne (now a country park) [TQ 6844 7016] and Paddlesworth [TQ 692 618], now a private fishing lake.

Chapter 7

Brickmaking

Brickmaking took place wherever there was brick earth and demand for bricks, but particularly in North Kent from the Medway Towns to Faversham, which had access to cheap water transport by barge to London. Many of the brick fields had no fixed plant such as kilns, instead using clamp burning, and have left little trace. Most brick fields with fixed plants, as at Burham, Lower Halstow, Otterham Quay, Murston Creek and Folkestone, have been demolished.

Extensive malm backs at the Burham Brick & Cement Company site [TQ 7210 1613]. Little remains of the kilns, but the gault clay pit [TQ 7220 1610] is used as a reservoir by Southern Water.

The kiln was erected sometime after 1871 at the South Eastern Railway Works at Ashford [TR 0162 4165] to make fire bricks for lining locomotive fire boxes. It was subsequently a store for acetylene, which was used in lighting carriages. Latterly it has been used as a café.

Hammill Brick Company works [TR 2939 5578] used old Goodnestone and Woodnesborough workshops. Later buildings have been cleared for redevelopment. The works were operated from 1923 to 2008, and obtained clay via a 2-foot tramway from a pit that can be seen 500 yards away, south of Woodnesborough Colliery.

Funton Brickworks [TQ875 677] were built in the late 1930s but did not begin production until after 1945. They are now closed and awaiting reuse.

Chapter 8

Hops, Malting and Brewing

Hops used in the brewing of beer have been grown on a commercial scale in Kent since at least the sixteenth century, with hop gardens widely distributed across the county. The oast house in which the hops were dried has become a symbol of Kent. The other essential building was the hop-pickers' hut that housed the annual influx of Londoners, who were the essential labour force before machine pickers.

Malted barley is the basic ingredient of ale, with the result that, to satisfy brewers, over 200 malthouses have operated in Kent. Over seventy malthouses survive, often converted and unrecognisable as maltings. Most are near the light loams of North Kent, which produce high barley yields, with the earliest examples dating from the fifteenth century and the last purpose-built malthouse from 1898.

Kent's once-thriving brewing industry was reduced from 100 breweries in 1880 to just one substantial firm, Shepherd Neame, by the 1990s. This resulted from changing consumption habits, restrictive licensing and commercial pressures, which lead to the amalgamation of Kent breweries with national brewers and their subsequent closure. Many breweries were situated near town centres and were difficult to convert to other uses, becoming prime redevelopment sites.

Little Golford Oast, Cranbrook [TQ 7998 3656], was constructed in around 1750 and only measures 21 feet by 15 feet. Originally it contained a brick furnace with a drying floor and a cooling floor section, but is currently converted for use as a garage.

Biddenden Oast [TQ 852 390] was derived from a mid-sixteenth-century barn, which was extended and fitted with an internal kiln in the late eighteenth century. It was converted into a house in 1980.

Eighteenth-century oasts with internal kilns are more typical of the Weald hop-growing area, like this example at Golford, near Cranbrook [TQ 7980 3651].

Sissinghurst Castle Oasts [TQ 8073 3838] has six 12-feet-square kilns, which were constructed in the late eighteenth century; two roundels were added in the third quarter of the nineteenth century. They last dried hops in 1967 and are now used as a display area by the National Trust.

The interior of a roundel at Sissinghurst Castle Oast, with a horse-hair mat on the drying floor, scuppet for moving hops, and a hop pocket into which the hops were pressed after drying.

A Weeks 'side winder' hop press at Sissinghurst.

Patrixbourne Oast [TR 1900 3524] was built by the Marquess of Conyngham in 1869. It is equipped with three brick roundels to the rear and a lucam over the road. It is now in residential use.

Nashenden Farm Oast, Nashenden Lane, Borstal [TQ 7317 6591], was built as a fire-proof oast using iron girders and joists, and concrete. Last used in 1981, it has been converted into two dwellings.

Beltring Hop Farm Oasts [TQ 6740 4750] has the largest concentration of oasts in Kent, consisting of four houses each with five brick roundels. They were built by hop farmer E. A. White towards the end of the nineteenth century and purchased by Whitbread Brewery in 1920. They are now used as a visitor attraction.

Oasts in East Kent appear to have favoured louvred vents, as in this example at Shalmsford Street [TR 092 548].

A traditional oast, with a modern oast behind, at Tutsham Farm, West Farleigh. Modern oasts were oil fired and usually had a long ridge cowl.

Hop-pickers' huts in School Lane, Wouldham, [TQ 7169 6427] now used as dog kennels.

Hop-pickers' huts, East Farleigh [TQ 734 534], associated with Court Lodge Farm.

Hollingbourne Malthouse [TQ 8446 5533] is a fifteenth-century, timber-framed building, which was in use by the Atkins family till the early 1900s.

Bredgar Malthouse [TQ 893 596] was a seventeenth-century malthouse with a square buttressed kiln. The thatch is a later addition.

Easole Street Malthouse, Nonington [TR 2628 5224], was built in 1704 with an internal kiln, thatched storage and growing area, and characteristic small windows.

The eighteenth-century malting, with four internal kilns, in West Street, West Malling [TQ 6779 5777], was converted into shops in the nineteenth century.

An eighteenth-century farm malting and hop oast at Court Farm, Stansted [TQ 608 621], with characteristic small windows.

Hythe Malthouse [TR 1595 3476], dating from the first half of the nineteenth century, was associated with Mackeson's Brewery, which stood adjacent to the west.

No. 2 and No. 3 Maltings, Hadlow [TQ 6320 4975], associated with the Close Brewery of Kenward and Court. No. 2 on the right was built in 1859, while No. 3 with the decorative gable end was built in 1880.

St Dunstan's Malthouse, Canterbury [TR 1433 5832], associated with Flint's St Dunstan's Brewery, dates from the third quarter of the nineteenth century. A second example has been demolished.

Russell's Maltings, West Street, Gravesend [TQ 6454 7446], were multi-storey maltings, with two kilns at the north end built in the 1880s.

St Stephen's Maltings, Canterbury [TR 1480 5870], was the last major malthouse built in Kent; it was constructed in 1898 for the Mackeson Brewery at Hythe.

Perry Street Oast, Faversham [TR 0097 6038], was purpose built as a hop oast, but doubled as a malting.

Knole House Brewhouse, Sevenoaks [TQ 54995424], serviced the requirements of the Knole estate from the sixteenth century, and is now a National Trust tearoom. Estate and farm brewhouses disappeared rapidly in the first half of the nineteenth century.

Fort Brewery, Margate [TR 3540 7117], operated by Webb & Co. from before 1873.

Hulke's Lion Brewery, High Street, Rochester [TQ 7514 6796], where the brewhouse dates from the 1780s.

Reeve's Brewery and tap, Margate [TR 3552 7100], has some surviving buildings through the tunnel. Brewing ceased in 1907.

Rigden's Brewery, Faversham [TR0174 6125], was remodelled in the late 1870s and early 1880s.

The tun house at Rigden's, Faversham.

The office and gate to Fremlin's Brewery, Earl Street, Maidstone [TQ 7582 5583]. The brewery was set up in 1790, acquired by Ralph Fremlin in 1861, modernised in the 1880s and closed for brewing in 1972 by Whitbread. The brewery site is now a shopping precinct.

St Dunstan's Brewery, Canterbury [TR 1433 5832], is on a brewery site dating from the eighteenth century; it was operated by Flint & Kingsford until 1923, and closed by Alfred Leney & Co. of Dover in 1929.

Troy Town Steam Brewery, Rochester [TQ 7425 6842] opened in 1750, and was modernised in around 1860 by Tomlyn Shepherd, after which it traded as Troy Town Steam Brewery. It was operated by the Woodhams family from 1865 to 1918, when it was purchased and closed by Style & Winch.

West Malling brewery [TQ 6826 5778] was built in 1880 and closed in 1939.

The office building in Dolphin Passage, Dover [TR 3204 4146], is all that remains of Alfred Leney's Phoenix Brewery, which, together with four malthouses, once dominated the centre of the town.

The Shepherd Neame Brewery in Court Street, Faversham [TR 0154 6155], was reputedly founded in 1698.The buildings generally date from the nineteenth century. Brewery tours are available.

Chapter 9

Water Mills

Water mills are widely distributed along Kent's rivers and streams. Initially used for grinding corn for flour, water mills have over the centuries been adapted to perform whatever was the most profitable function at the time, including fulling for the textile industry, paper making, oil-seed crushing, working bellows and forge hammers in iron works, working incorporating mills in gunpowder works, and even as saw mills. The number of active mills was recorded as 351 in the Domesday survey, but had fallen to 136 according to Coles Finch between 1819 and 1843; the number thereafter declined rapidly with the application of steam power and with the milling of imported grain in steam-powered mills at the ports. Surviving rural mills have mostly become private houses and are not accessible, with only Crabble Mill at Dover open to the public and grinding corn.

Farningham Mill [TQ 5451 6700] was built by Charles Colyer in the late eighteenth century to replace a corn mill that had stood at the location since 1610. The Darent had supported mills, many of which have been demolished, along its length from Westerham to Dartford, some of which were also used for fulling, papermaking and gunpowder.

Tonge Mill [TQ9344 6356] was built in 1837, and the mill house in 1866. A steam engine was added later with a square chimney of 100 feet, together with a bakery.

A mill at Chegworth [TQ 8501 5269] occupied this site in the late seventeenth century. The lower storey is built of brick and ragstone from the late eighteenth/early nineteenth centuries, with two upper storeys of weatherboard. The machinery of the mill is complete, including three pairs of stones. It continued to grind corn until the late 1960s, making it the last commercially working watermill in Kent. It is a survivor of approximately thirty mills that once operated on the river Len.

Chegworth Mill's iron pen trough and overshot wheel were supplied by William Weeks, engineer and iron founder of Maidstone.

French Mill, Chilham [TQ 077 534], built *c.* 1800. Originally, an iron breast shot wheel inside the mill worked six pairs of stones on the second floor. It was converted to steam in the nineteenth century when the mill pond was infilled. The boiler house and chimney have been demolished.

Evegate Mill, Smeeth [TR 0637 3808]. The mill house dates from the early eighteenth century and appears to have been the original mill with the mill race running through an arch at the east end. The present mill was built in 1862, when the water was diverted to an overshot wheel at the rear.

Edenbridge Mill, 87 High Street [TQ 4442 4606], dates from the eighteenth century. Standing some distance from the River Eden, the mill's undershot wheel was powered from a leat running under the road and into the building from a connection with the Kent Brook, several hundred yards to the south-west. Adjoining to the left is the mill house, of early nineteenth-century date.

Wickhambreaux Mill on the Little Stour [TR 215 581] is a five-storey mill, built in the early nineteenth century and worked until 1955. The cast-iron breast shot wheel drove five pairs of stones, latterly producing animal feed. The mill has been converted into flats.

Crabble Mill, Dover [TR 2975 4318], was built as a corn mill in 1812 to grind flour for troops in the area. It worked five pairs of stones until 1890.

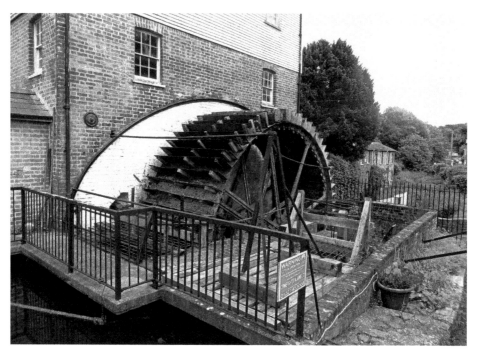

The breast shot waterwheel at Crabble Mill.

Hanover Mill, Mersham [TR 050 391], was built by millwrights Holman Brothers in 1879 as a rebuild of an eighteenth-century mill, using its ragstone foundations. It was equipped with a 12-foot breast shot wheel and three pairs of stones.

Belvedere Mill on Faversham Creek [TR 0163 6167] is a nineteenth-century, steam-powered corn mill. It was able to receive imported grain by barge and has a lucam over the creek. The boiler house, chimney and two external silos, which stood against the north wall, have been removed. The building is now apartments, with a restaurant filling the ground floor.

H. S. Pledge & Sons Ltd, East Hill Mill, Ashford [TR 0153 4281], was built by Henry Pledge in 1901 as a water and steam mill. The mill ceased work and was sold to become a nightclub in 1972.

Chapter 10

Windmills

According to Coles Finch, there were 134 windmills in the period 1903–10, many not working and their associated bakeries gone. By 1930 there were just two post mills, two tower mills and thirteen smock mills active, with another fifty-three derelict or converted. This number has fallen to fourteen complete windmills in preservation, of which eight (Chillenden, Stelling Minnis, Margate, Herne, Meopham, Wittersham, Cranbrook and West Kingsdown) are owned by Kent County Council and open to the public, and twelve that are capless stumps.

West Kingsdown Mill [TQ 5813 6219], a tarred smock mill, was originally erected at Farningham in 1800 and was only transported to its present location in 1880. It was operated by the Norton family until 1909.

Union Mill, Cranbrook, [TQ 7791 3590] was built in 1814 and worked until 1957. It was raised onto a three-storey brick base to become the tallest smock mill in Kent, at 70 feet. It was modernised in 1840 with a fantail, new wind shaft and patent sweeps. It had an auxiliary steam engine in 1863, a gas engine in 1919 and finally electric power in 1954, working three pairs of stones. It retains its machinery in working condition.

Herne Mill [TR 18496650] was built as a smock mill in 1789. To catch the breeze, it was lifted onto a brick base in 1858. In modern times, it used a tractor to provide power when the wind was calm, and remained working until 1980, retaining its original machinery.

Killick Bros, millwrights, built the small Meopham Mill [TQ 6390 6520] in 1821 as a 'model' to demonstrate their mill-building capabilities. Of timber on a brick base, it ran until 1959 using electric power.

Woodchurch Mill [TQ 943 353] was a late eighteenth-century smock mill, moved to its present site from Susan's Hill in 1820. It ceased work in 1926, but has been saved by the Friends of Woodchurch Windmill and Ashford Borough Council. It is open to the public on summer weekends.

Chillenden Mill [TR 2689 5430], built in 1868, is a post mill, constructed in timber with an open trestle base and a tail pole for turning the mill. It was in use working two pairs of stones until 1949.

Willesborough Mill [TR 0312 4214] is a five-storey smock mill, built in 1869 with a mill house adjoining. It later had an auxiliary oil engine, which is currently a 1906 Hornsby gas engine. Restored by Ashford Borough Council in 1991, it is open on summer weekends.

Sarre Mill [TR 259 651] was built in 1820 by Holman and worked until 1940 using a gas engine. Restoration was completed in 1991. It is open to the public, with a tearoom and stone-ground flour sales. Other mills open on summer weekends include: Drapers Mill, Margate [TR 362 700], which was built 1845 by Holmans and worked until the 1930s, and now run by a Crossley gas engine for demonstrations; Davison's Mill, Stelling Minnis [TR 146 466], built in 1866 and worked until 1970, and Stocks Mill, Wittersham [TQ 913 273], which is a post mill built in 1781.

Black Mill, Borstal Hill, Whitstable [TR 105 652], is a five-storey smock mill, built in 1792 and worked until 1906. It has since been the home of Henry Irving, the actor; a motel; and is currently part of a residential development.

Rolvenden post mill [TQ 838 315] dates from *c.* 1775 and worked till 1882. The fabric has been restored by the Hole Park estate.

Chapter 11

Oil-Seed Mills

Mills to crush locally grown linseed and rapeseed to produce oil and cattle cake were located mostly by the Medway between Gillingham and East Peckham, but also at Conyer Quay and Dover.

Tutsham Mill, Teston [TQ 7088 5301], is said to have been built to a design by John Rennie in 1808. It was powered by a turbine, using the fall in the river at Teston Lock on the Upper Medway Navigation, with the outfall to the left of the picture. It came into the hands of Thomas Boorman in 1847 and ran as Boorman Wild & Co. till bankruptcy in 1862. Roger Leigh of Barham Court installed Anglo Machinery and leased the mill to Stewart Brothers & Spencer after 1884, only for the mill to be burnt out in 1889.

The works at Cuxton [TQ 7134 6660] were used by W. J. Mackay after their oilseed mill in Gillingham was burnt down in around 1895; they were later occupied until 1971 by the British Bestos & Basket Co. to make fruit baskets and trugs.

Chapter 12

Engineering and Foundries

In the nineteenth century, engineering workshops and iron foundries developed to meet the specific needs of Kent's agriculture and growing industries in towns such as Ashford, Dartford, Maidstone and Rochester. Many familiar firms are now only known through the survival of their products, with little trace left of their factories, which were often situated near town centres and have become valuable brown field sites for redevelopment. These include:

- Aveling & Porter of Rochester and their successor firm (after 1934) Wingets.
- William Weeks of the Perseverance Ironworks, Waterside, Maidstone, known for crop sprayers, hop presses, towed scarifiers and in around 1920 the Weeks-Dungey tractor.
- Drake & Muirhead, which started up in 1882, later to become Drake & Fletcher of West Borough, Maidstone, also manufactured crop sprayers and, in 1903, a three-cylinder petrol driven tractor.
- Halls of Dartford, where Trevithick was once employed, who were mill wrights, engineers and manufacturers of machinery, including steam engines; equipment for the paper industry, including a papermaking machine by Brian Donkin; and equipment for the gunpowder industry. Latterly, they specialised in lifts and a refrigeration plant. The extensive factory site at Dartford has been redeveloped.
- Collis & Stace of Strood and Taylor & Neates of Rochester, who manufactured equipment for the brick, cement and barge-building industries.
- Norman bicycles and motorcycles, manufactured in Beaver Road, Ashford, until closure in 1961.
- The aircraft industry in the shape of Short Brothers, The Esplanade, Rochester, from 1914 until 1948.
- Perceval Aircraft at Gravesend Airport between 1934 and 1936.

However, a range of factory sites have managed to survive.

W. A. Stevens Ltd was established in St Peter's Street, Maidstone [TQ 7551 5610], in 1897 as electrical engineers making dynamos and electric motors. In 1906 Stevens experimented with electric motors fitted to a petrol engine as a means of powering commercial vehicles, doing away with the need for a gearbox. He went on to use the Hallford bus chassis to build petrol-electric buses.

Pictured is the interior of the Tilling Stevens factory, which is now used as a timber store. It produced road vehicles, particularly buses after the firm was taken over by Thomas Tilling in 1913. The firm continued to make buses, vans and trucks until it merged with the Rootes Group in 1950.

Tilling Stevens [TQ 7559 5610] factory, built in 1917. Designed by Wallis Gilbert & Partners, using reinforced concrete and glass, it is an early example of an Art Deco building.

Len Engineering, Palace Street, Maidstone [TQ 7600 5555], was built on the site of Maidstone tannery. In 1917 William Rootes used the site for the repair of aero engines before switching to vehicle production until 1937. In 1938/39, the factory was rebuilt as a model car showroom and workshop.

Hallford bus, built by Halls of Dartford.

Weeks Ironworks towed scarifier at the Brattle Farm Museum, Staplehurst.

The former workshop of Wingham Engineering in Goodnestone Road, Wingham [TR 2462 5702], one of several steam contractors in Kent who used Aveling & Porter engines and carried out their own maintenance.

Buildings dating from
1847 at the South
Eastern Railway Works,
Ashford [TR 0180 4160],
which incorporated
the locomotive hall,
with smiths' shops and
machine shops on the
southern side.

The locomotive hall
interior at Ashford
Railway Works.

The carriage shop at the
South Eastern Railway
Works, Ashford [TR 0153
4180], built in 1858. The
Italianate water tower
was added in 1898,
when the building was
converted into a saw mill
after new carriage works,
the Klondyke Works (now
demolished) were built on
the opposite side of New
Town Road. Many other
buildings on site have
been demolished.

The gate lodge to the Southern Railway Works [TR 0158 4168], built in about 1850. The Italianate clock tower is a later addition and was probably built to complement the carriage-works tower.

The smithery at Chatham Dockyard [TQ 7601 6931] was built from 1806 to 1808 as three ranges around an open court, with two lodges that served as offices. The courtyard was roofed in around 1860. The smithery was used for forging the increasing number of iron components used in warships after Seppings introduced iron knees in 1806. Little original equipment remains, with the building used as a gallery for model boats and as an exhibition space.

The Factory Building at Chatham Dockyard [TQ 7637 6995] originated as a covered slip at Woolwich Dockyard; it was dismantled and re-erected at Chatham as an economy measure. Dismantled covered slips and other iron buildings re-erected at Chatham served as the engineering factory, foundry, carpenters' shops and workshops. The building is seen stripped of its cladding before being converted into a retail outlet.

Rochester Airport [TQ 7432 6498] in 2015. Top centre with the rounded roofs are the 1934 hangars used by Short Bros as an assembly plant, including for the Stirling bomber. The tower blocks behind were erected by Elliott Automation when they used the site as an avionics factory. Currently the plant is occupied by BAE Systems. The shed bottom, left, was built in 1938 as a workshop for an Elementary Flying Training School. Short Bros had a satellite factory in Knight Road, Strood [TQ 7310 6854], built in 1938 for the production of wings and tail planes for the Stirling bomber and Sunderland flying boat. It is now used as a transport depot. (Kelvin Carr, Airport Manager)

Chapter 13

Tanneries

In the nineteenth century, tanneries occurred mainly in urban centres; these included Ashford (dating from the early 1700s), Canterbury (where there had been a tannery at the St Mildred's site from Roman times), Dartford (built in 1858 and closed in 1938), Dover, Edenbridge, Maidstone, Milton by Sittingbourne, Sandwich (in Loop Street from 1832 and closed 1940), Southborough, Tenterden, Tunbridge Wells, and West Malling, with a few in rural locations such as East Peckham and Shorne (late 1600s to early 1900s). The number of tanneries had fallen to nine, as listed in Kelly 1913. The decline in numbers continued with the last closure, St Mildred's Canterbury, occurring in 2002. Most tannery buildings have been demolished and the sites redeveloped.

The tanner's house at 92 High Street, Edenbridge, [TQ 4439 4604], was one of a range of fifteenth-century houses associated with Edenbridge tannery, which stood behind. From around 1860 the tannery was operated by the Whitmore family, who had formerly operated tanneries at Westerham. The site, closed as a tannery in the 1970s, is now occupied by a supermarket; the only traces that remain are a memorial to workers killed in the First World War on the wall of 94 High Street, and the leat to Edenbridge mill.

The Edenbridge Tannery memorial plaque, erected in 1922 to workers killed in the First World War; it includes Capt. R Whitmore, the son of one of the directors.

The leat, which had been dug from the Kent Brook to Edenbridge corn mill, ran through the tannery, providing water to wash hides.

Tannery building at Tanyard Farm, Lenham [TQ 9026 5193]. Here water from the Stour was diverted into the building via the arch at the left-hand corner and flowed through at basement level, providing a washing facility for hides.

One building remains of the once-extensive St Mildred's Tannery, Stour Street, Canterbury [TR 1456 5760]. The front of the building with loading doors was a leather warehouse, and the rear was for hide drying; it once had shuttered windows.

Chapter 14

Coal Mines

Coal was discovered at a depth of 300 metres at the Channel Tunnel workings at Shakespeare Cliff [TR 295 393], 3 miles west of Dover, in 1890, resulting in the Kent Coalfields Syndicate Ltd digging shafts in 1896. Constantly hampered by water, as flooding was a problem for the coalfield, coal was not hit until 1903, with little extracted before closure in 1915.

The Dover mine lead speculative companies to sink over forty boreholes in East Kent to find the extent of the coalfield and possible workable seams. A leading figure in promoting coal companies was Arthur Burr, who started five collieries between 1904 and 1910, and who by 1910 was associated with twenty-two interlinked companies, which acquired the mineral rights to most of the coalfield. Lacking the capital required to exploit the concessions, Burr tried to work with cheap, second-hand equipment, leading to early liquidation and abandoned projects.

Four pits survived to produce any quantity of coal. Tilmanstone [TR 288 505] was started by Burr in 1906, plagued by flooding, but hit commercial quantities of coal in 1913, producing 150,000 tons. In receivership in 1914 and 1926, the colliery was bought by Richard Tilden Smith and the Tilmanstone (Kent) Colliery Co., finally passing to the NCB, who closed it in 1986. The mine was always uneconomic. The site is being redeveloped, with the most significant legacy being 230 miners' houses built at Elvington. Snowdown Colliery [TR 247 512] was initiated in 1907 by Arthur Burr with three shafts, but flooding delayed the production of coal until November 1912. The colliery was sold to Pearson Dorman Long in 1924, who built the village of Aylesham after 1925 for 650 workers' families, with shops, club and a school. Under the NCB, the colliery closed in 1987. Chislet Colliery [TR 209 619] was begun in 1918 by the Anglo Westphalia Kent Coalfield Ltd, using the cementation process to seal the shaft against water. Reorganised as Chislet Colliery Ltd in 1924, with 165 miners' houses being built at Hersden, the mine always struggled financially. It was modernised by the NCB in 1947 but finally closed in 1967 and demolished. Betteshanger Colliery [TR 336 530] was opened in 1924 by Pearson & Dorman Long (who had also purchased Snowdown, Wingham and Woodnesborough collieries) and produced coal from 1927. The mine was closed in 1989, and the site is being redeveloped.

Failed projects that have left remains include a site in Hammill Road, Woodnesborough [TR 2935 5575], which was begun in 1910 under the title of Goodneston & Woodnesborough Collieries Ltd. Some surface buildings were erected and provision was made for two shafts, but the project was mothballed in 1914 and sold to Pearson & Dorman Long in 1923. The chimney and most buildings were demolished before the site was sold on to Hammill Brick Co. for brickmaking, which ceased in 2006. Wingham Colliery in Staple Road [TR253 569] started in 1911, but hit water and was unable to afford pumps; the site was mothballed until sold to Pearson & Dorman Long, who did no further work. The site was purchased by Wingham Engineering Co. in 1934, who demolished the chimney. In 1947 the site passed to Grain Harvesters Ltd, who incorporated the former winding house into a grain store. Stonehall Colliery, Lydden [TR 271 456] saw work started on digging three shafts in 1913; however, this stopped in 1914 when water was hit. Work resumed in 1919, but the colliery was abandoned in 1921, before the coal seams were reached. Guilford Colliery, Coldred [TR 280 469], was started by Arthur Burr in 1906 with three shafts that reached 1,272 feet, before water was hit and work stopped in 1910. The site was sold to a French company in 1919, who tried to seal the shafts using the cementation process without success, giving up in 1921. The explosives magazine and workshops have been converted to residential uses.

A shallow lignite mine was opened in 1947 by the Cobham (Kent) Mining Co. Ltd. to exploit a 6-foot seam on the Darnley estate [TQ 674 696]; however, due to problems with water, roof falls and the poor quality of the product, the scheme was abandoned in 1953. The new carriageway of the A2 wiped out all trace of it, but subsidence shows up as depressions in nearby woods.

Snowdown Colliery buildings. The head stock has been removed, but the winding house and workshops remain.

Hammill Colliery workshop buildings.

The workshop at Stonehall Colliery, Lydden.

Chapter 15

Warehouses

Warehouses were built for the storage and protection of agricultural or manufactured products and for the collection, onward transit and distribution of goods. These range from tithe barns and granaries to cloth halls, manufacturers' stores, carriers' depositories and timber stores.

Lenham tithe barn [TQ 8988 5205] was built in the late fourteenth century by St Augustine's Abbey, Canterbury. It was extended in the early sixteenth century and received a tiled canopy over the door in the nineteenth.

Westenhanger Castle barn [TR 1224 3719] is L shaped, with stables built *c.* 1575.

The Town Warehouse, on the wharf at Conduit Street, Faversham [TR 0158 6161], was constructed *c.* 1475. It is currently in use by Sea Scouts.

The granary and warehouse at Standard Quay, Faversham [TR 0196 6194], dates from the late seventeenth century. It was built from timbers from Faversham Abbey refectory after its demolition in 1671.

The fitted rigging house and stores on Anchor Wharf, Chatham Dockyard [TQ 7578 6898], is the largest naval storehouse at 700 feet, completed in 1785.

Pickford's Depository, opposite Canterbury East station [TR 1470 5730], was built in the late nineteenth century as a transit store.

Oyster Bay House, Iron Wharf, Faversham [TR 0212 6203], was built in the mid-nineteenth century as a three-storey granary, capable of receiving cargoes from the newly improved Faversham Creek.

Timber storage sheds at Swan Quay, Faversham [TR 0160 6164], a class of building that has largely disappeared.

Horton Kirby paper warehouse built *c.* 1880.

Chapter 16

Utilities

The water supply became a major issue in the nineteenth century, driven by concerns about public health, leading to provision by private companies and municipal undertakings. Some older works remain, despite modernisation and rationalisation of the industry.

Town gasworks were part of the townscape from the 1820s, but became redundant with the change to North Sea gas. Most buildings have been demolished, although a few gasometers continue to be used for storage or are awaiting demolition, including: Brielle Way, Queenborough; Jackson Road, Canterbury; Sea Street, Herne Bay; Golf Road, Deal; Medway Wharf Road, Tonbridge and Gravesend Canal Basin. Electric power stations, including those built post-1945, are similarly victims of rationalisation, although a few earlier buildings remain.

Cobham village pump [TQ 6693 6847] stands over a tank, which collects spring water. The plaque on the wall reads, 'This reservoir was enlarged and rebuilt with the waterworks hereto belonging at the sole expence of the Right Honourable John Stuart Earl of Darnley and given to the parishioners to commemorate his coming of age April the 16th 1848'.

The water pump from Dickens' residence, Gadshill House, Higham, which dates from 1857, was worked by a donkey. It is currently in the garden of Eastgate House, Rochester [TQ 7448 6833].

The town reservoir on Tanners Hill, Hythe [TR 1641 3503], was constructed in 1868 by Folkestone Waterworks Company. The site was occupied by a tannery, with a tan pit being redeveloped as a reservoir, a 3-metre-deep tank fed by nearby springs.

Copton waterworks, Faversham [TR 0129 5954], was built in 1863 by Faversham Water Company. The complex consisted of a wind pump in the form of a tower mill, from which the sweeps were removed in 1930; an engine house built in 1914; a workshop; an office; and workers' cottages.

East Farleigh pumping station [TQ 7433 5355], built in 1860. Designed by James Pilbrow in Egyptian style for Maidstone Water Company, it pumped water collected in a reservoir from springs to a reservoir at Barming, from where the water could flow by gravity (over a 300-foot fall) to Maidstone.

East Farleigh pump house, built in 1878 to increase capacity. The works were shut in 1897 after a typhoid outbreak in Maidstone.

Dover Waterworks, Connaught Road [TR 3218 4213], built in the late nineteenth century in an Egyptian-revival style with decorative windows. The height was originally necessary to accommodate beam engines, a practice continued here despite the use of triple-expansion engines. The boiler house and chimney have been removed. The site is still operational.

Wingham Waterworks
[TR 2431 5531] was built
in 1903–5 for Margate
Waterworks with a mixture
of Baroque and Art
Nouveau styles, as seen
in the door pediment and
stained-glass windows.

Boxley Waterworks,
The Street, Boxley
[TQ 774 594], together
with a reservoir behind,
was constructed in 1939
to supplement Maidstone's
water supply.

Newnham pumping
station [TQ 9749
5968] was built by Mid
Kent Water Company
between 1936 and 1937.
Integral to the scheme
was a 100,000-gallon,
concrete-lined reservoir and
a water softener.

The Brook pumping station, Chatham [TQ 7598 6782], was built in 1925 and operational by 1928, pumping storm water and sewage to Motney Hill disposal works.

Gillingham gasometer [TQ 7826 7826].

The gasometer in Victoria Road, Dartford [TQ 5415 7456], the survivor of three dating from the early 1900s.

The purifier house at Faversham Gas Light & Coke Company [TR 0129 6155] was built in 1900 but never used for purifying. Now in Morrisons car park, it has been renovated by the Faversham Creek Trust.

Generating station by Faversham Creek, built in 1911 [TR 0159 6162].

Tonbridge power station, The Slade [TQ 5885 4670], was built in 1902 in the Arts and Crafts style, influenced by Voysey. The front was a two-storey office with a single-storey power house at the rear. The station generated power until 1951, after which machinery was removed and the boiler house and chimney demolished.

Kent Electric Power Co. Ltd, Frindsbury [TQ 7448 6955], was built *c.* 1904 and supplied power to the Medway Steel Company at Strood, which had an electric furnace.

Dungeness A and B nuclear power stations [TR 0832 1695]. Dungeness A started generating in 1965 and ceased in 2006; it is currently being decommissioned. Dungeness B started generating in 1983 and is expected to run until 2028. There is a visitor centre.

Bibliography

Armstrong, A., (Ed), *The Economy of Kent 1640–1914* (Boydell, 1995).

Banbury, P., *Shipbuilders of the Thames and Medway* (Newton Abbott: David & Charles, 1971).

Barber, N., *A Century of British Brewers 1890–1990* (Brewery History Society, 1996).

Barker, T. C., *Shepherd Neame* (Cambridge, 1998).

Barnes and Innes, *A Nineenth Century Cement Works at Cliffe* (Kent Archaeological Review, 1984).

Bennett, C. E., *The Watermills of Kent, East of the Medway* (Industrial Archaeology Review, Vol I, No. 3, Summer 1977).

Chalklin, C. W., *Seventeenth Century Kent: A Social and Economic History* (Longman, 1965).

Cleere, Henry, and Crossley, David, *The Iron Industry of the Weald* (1995).

Coad, Jonathon, *Historic Architecture of Chatham Dockyard* (National Maritime Museum, 1982).

Coles Finch, William, *Watermills and Windmills* (Sheerness: Arthur Cassell, 1976).

English Heritage, *Research Department Report, 11-2011 'Cliffe Explosive Works'*.

Filmer, Richard, *Hops and Hop Picking* (Princes Kisborough: Shire, 1998).

Francis, A. J., *The Cement Industry 1796–1914* (Newton Abbott, David & Charles, 1977).

Fuller, Michael, *The Watermills of the Malling and Wateringbury Stream* (Maidstone: Christine Swift, 1980).

Furnell, K. J., *Snodland Paper Mill* (Snodland: Townsend Hook, 1980).

Hilton, J. A., *History of the Kent Coalfield* (1986).

Lawson, Terence, and Killingray, David, (Eds), *An Historical Atlas of Kent* (Chichester: Phillimore, 2004).

Miller, H., *Halls of Dartford 1785–1985* (London: Hutchinson Benham, 1985).

Percival, Arthur, *The Faversham Gunpowder Industry and its Development* (Faversham: The Faversham Society, 1986).

Perks, R. H., *George Bargebrick Esquire* (Rainham, Meresborough, 1981).

Pike, Geoffrey, *Copperas and the Castle* (Whitstable: Friends of Whitstable Museum, 2000).

Philp, Brian, *The Dartford Gunpowder Mills* (Council for Kentish Archaeology, 1984).

Preston, J. M., *Aveling and Porter* (Rochester: North Kent Books, 1987).

Preston, J. M., *Industrial Medway* (Rochester: Author, 1977).

Preston, J. M., *Malting and Malthouses in Kent* (Stroud: Amberley, 2015).

Rowley, Chris, *The Lost Powder Mills of Leigh* (Leigh Historical Society, 2008).

Salmon, Anne, *A Sideways Launch* (Rainham: Meresborough Books, 1992).

Sattin, D., *Barge Building and Barge Builders of the Swale* (Rainham: Meresborough, 1990).

Shorter, A. H., *Papermaking in the British Isles* (Newton Abbott, David & Charles, 1971).

Spain, R., *The Loose Watermills I* (Archaeologia Cantiana, Vol. 87, 1972).

Spain, R., *The Loose Watermills II* (Archaeologia Cantiana, Vol. 88, 1973).

Straker, Ernest, *Wealden Iron* (Newton Abbott: David & Charles reprint, 1969).

Taylor, Michael, *Shorts* (London: Jane's Publishing, 1984).

Tonbridge Historical Society, *Tonbridge's Industrial Heritage* (Tonbridge, 2005).

Turner, G., *Ashford and the Coming of the Railways* (Christine Swift, 1984).

Walton, R. and I., *Kentish Oasts, 16th–20th Century, Their History, Construction and Equipment* (Egerton: Christine Swift, 1998).

West, Jenny, *The Windmills of Kent* (London: Skilton and Shaw, 1979).

Willmott, Frank, *Bricks and Brickies* (Rainham: Author, 1972).

Willmott, Frank, *Cement, Mud and Muddies* (Rainham: Meresborough, 1977).

Useful, but out-of-date guides:

Eve, David, *A Guide to the Industrial Archaeology of Kent* (Association for Industrial Archaeology, 1999).

Haselfoot, A. J., *The Industrial Archaeology of South-East England* (Batsford, 1978).